LLOYD HERVEY II

Political Design

Individual Values, Team Dynamics, Industry Culture

EXIT 127

First published by EXIT 127 2026

First edition

ISBN: 979-8-218-93123-0

This book was professionally typeset on Reedsy.
Find out more at reedsy.com

Contents

Introduction   1

Chapter 1: What is Political Design?   2

Chapter 2: Individual Values   5

Chapter 3: Team Dynamics   8

Chapter 4: Industry Culture   11

Chapter 5: The Forces That Shape Political Design   15

Chapter 6: Real-World Applications of Political Design   19

Chapter 7: Designing for Liberation   27

Conclusion   30

Appendix A   32

Appendix B   35

Appendix C   38

Appendix D   41

About the Author   45

Introduction

Introduction: Political Design

Every product, platform, and policy we interact with is not just the result of creativity and technical expertise, it's shaped by dynamics purely beyond design principles. **Political Design** is the unseen framework that determines what gets built, how it functions, and who benefits from it. It exists at the intersection of **Individual Values, Team Dynamics, and Industry Culture,** where ideas, collective agendas, and industry structures dictate the evolution of technology, business, and daily life.

Political Design influences everything from the Apps we use to the way financial systems operate. Additionally, the decisions made whether through **power, influence, or access** will have a lasting impact on social progression. Understanding these dynamics gives us insight into why some ideas flourish while others diminish. These dynamics also demonstrate how individuals, teams, and industries directly contribute to the design of our world.

This book explores the components of Political Design, breaking down its fundamental elements and real-world applications. By the end of this journey, we'll recognize the specific dynamics at play behind the structures we take for granted and how we can navigate, challenge, and shape Political Design to impact future generations.

Chapter 1: What is Political Design?

Individual Values. Team Dynamics. Industry Culture.
Understanding how individual influences, group structures, and regulatory frameworks shape design decisions.

Political Design is the process by which individual values, team dynamics, and industry culture shape the development and evolution of technology and user experiences. It's how ideas, interactions, and rules determine what gets made and how people use it.

Unlike traditional design, which prioritizes optimal user experiences based on fundamental design principles; Political Design recognizes that external factors such as personal beliefs, team decision-making, and industry norms actively shape the design process.

In today's world, every product, platform, and policy is created within a framework of political influences. These influences aren't always direct but often subtle emerging from the values and priorities of the individuals involved, the teams they represent, and the systems that govern.

Why Political Design Matters

Political Design matters because it defines the stakes of many decisions within a project experience. Whether those decisions are about features prioritized for a tech product, the structure of financial systems, or how health policies affect a population. These decisions impact who gets to innovate and who has the opportunity to participate in those innovations.

Consider the development of a social media platform. The decisions made in designing that platform such as how to moderate content, how to prioritize user data, and what features to highlight are shaped by the political dynamics behind it. Those with power in the company dictate how these decisions are made while external influence, such as public opinion or regulatory pressure, play a crucial role in defining what the platform becomes.

The same applies to the rise of cryptocurrency. The design of decentralized financial systems wasn't just about creating an innovative payment method. It involved decisions about how the technology would be regulated, who would have access to it, and how the wider financial system would adapt. All these decisions are deeply rooted in Political Design.

Political Design in Action

An example of Political Design in action is the smartphone evolution. The smartphone isn't just about a device that connects people. It's about who has the ability to implement these devices, who has access to the technology, and how industry leaders make decisions on data transparency. Companies like Apple™ and Google™ don't just compete for market share; they're also making decisions that influence societal norms, cultural shifts, and political behavior. Decisions on privacy, access to Apps, and data usage are all products of Political Design.

Consider also how AI technologies are evolving. The design of AI systems not only involves machine learning and technical algorithms; but also political decisions on who develops the technology, how it's governed, and what ethical considerations should be implemented. The people in charge of these decisions hold the power to shape the future of industries, economies, and society. As technologies evolve, the role of Political Design will become much more relevant.

The Role of Design Thinking in Political Design

In traditional design thinking, the focus is on creating innovative solutions that are practical, easy on the eyes, and user-centered. Political Design takes these principles a step further by considering how individual values, team dynamics, and industry culture shape solutions. Political Design demands conscious leadership to consider who will benefit from a particular solution, who may be harmed, and who may be left out.

For example, designing a digital platform for education isn't just about establishing engaging learning experiences. It's about who gets to decide what content is featured, who has access, and who governs the platform's purpose and educational policies. Recognizing these dynamics helps us make informed choices, challenge systemic obstacles, and design a world that is more authentic, transparent, and compassionate.

Political Design is everywhere, often working behind the scenes to influence what specific technologies, policies, and products become reality. These political dynamics give us a better sense of how society, technology, and culture will be impacted moving forward.

Chapter 2: Individual Values

How Personal Beliefs Affect Innovation
"The smallest action can lead to the greatest impact."

In the realm of Political Design, personal beliefs shape far more than individual behavior; they influence entire industries, impact technological innovations, and drive the outcomes of design decisions. Whether we're aware of it or not, our personal values; the things we hold dear, the beliefs we stand by, and the way we see the world; are at the heart of every choice we make.

Individual values play a pivotal role in determining how innovation unfolds. Every designer, developer, policymaker, and entrepreneur brings a unique set of experiences, beliefs, and perspectives to the table. These values can either promote or hinder progress, depending on the context in which they are applied. Whether it's deciding which features to include in a new tech product, choosing the ethical framework for an AI system, or determining the focus of a policy initiative, individual values impact the process.

The Power of Personal Beliefs in Shaping Design

When we talk about individual values in Political Design, we are really discussing how personal beliefs guide decisions and innovation. These beliefs don't just dictate how people interact with technology; they often determine what technologies are eventually established.

For example, the founders and decision-makers behind social media platforms didn't just create a product; they built ecosystems that reflect their views on connection, communication, and information sharing.

Facebook's early days, for example, were shaped by Mark Zuckerberg's belief in connecting people across the world. As the platform evolved, it reflected his personal values of transparency and openness, which were core to the early iteration of Facebook™. Over time, however, concerns around privacy and data sharing arose, forcing changes to how Facebook™ operated. This involved shifting public values and societal debates around ethical design.

Similarly, when developing cryptocurrency platforms, the underlying belief system was deeply embedded in the concept of decentralization. This idea promotes a distrust of traditional financial systems and the belief that control over one's money should lie with the individual, not institutions. This personal perspective in financial freedom and autonomy gave rise to a whole new industry and sparked innovations that continue to evolve today.

At the heart of individual values is an element of morality. What designers and creators believe is "right" or "wrong" can dramatically influence the solutions they develop. In many cases, the success or failure of an innovation isn't determined purely by its technical functionality or market demand; it's also deeply rooted in how aligned the product is with the creator's moral perspective.

Balancing Personal Beliefs and Public Interests

While personal values can drive innovation forward, there is a constant tension between individual interests and public good. Sometimes, what one person believes is a necessary innovation can conflict with what the market requires or the public expects. This tension is where Political Design becomes essential because it allows for the negotiation of these interests by balancing personal values against the broader social, cultural, and regulatory dynamics.

This is why personal beliefs are not always universally embraced. What one person sees as a force for good, another might perceive as a source of harm The challenge becomes how to design solutions that not only align with individual values, but also resonate with the needs, norms, and expectations of society as a whole.

Chapter 3: Team Dynamics

Why Collaboration, Conflict, & Leadership Styles Shape Outcomes
"Working together creates unique experiences that influence the world"

In the world of Political Design, it's not just about individual values; it's also about the collective energy and interplay of teams. The way people collaborate, communicate, and navigate the complexities of group decision making can have a profound impact on the design process. Relationships, communication patterns, and group structures often dictate the success or failure of innovation.

While collaboration can accelerate progress and spark creativity, conflict is also an inevitable part of the process. It's through navigating both harmony and tension that the most impactful designs come to life. The truth is, Political Design doesn't just arise from individual minds or top-down authority; it is the result of how teams manage competing ideas, confront obstacles, and leverage diverse perspectives.

The Role of Team Dynamics in Political Design

Team dynamics are the invisible forces that determine how teams operate, collaborate, communicate, and innovate. These dynamics can be affected by collaboration, conflicts, and leadership styles.

Collaboration is often seen as the ideal space where team members come together to brainstorm, exchange ideas, and find solutions. However, in Political Design, collaboration goes beyond cooperation. It's about combining the unique skills, values, and perspectives that team members bring collectively. The more diverse the team, the greater the potential for innovative, multi-dimensional solutions.

Consider the development of AI technology platforms and initiatives. These projects often bring together software engineers, data scientists, and business leaders. Each of these team members have their own perspective on the product's purpose, its ethical implications, and its potential impact on society. The strength of these teams lies in how well they manage to integrate and align their varying values and expertise into a cohesive solution.

Conflict also plays an essential role in Political Design. It's in the friction between opposing ideas and values that many of the most groundbreaking visions are born. Disagreements within teams are not simply moments of frustration, but opportunities to rethink, refine, and challenge existing assumptions. Conflict, when handled productively, forces teams to confront their biases, stretch their thinking, and ultimately come up with more well-rounded, resilient designs.

For example, when designing a new product or platform, the sales team might be focused on user acquisition and scalability. However, the engineering team might prioritize security and performance. This tension between priorities forces the teams to engage in dialogue, work through trade-offs, and innovate solutions that balance competing demands. As the design process moves forward, the final product may be far more refined, balanced, and effective than if everyone had agreed from the beginning. Conflict forces teams to engage in critical and often divisive issues, ultimately leading to more responsible and thoughtful innovations.

Leadership styles also play a significant role in determining how well collaboration and conflict are managed. A strong, empathetic leader can guide the team through moments of tension, ensuring all voices are heard, and preventing conflicts from escalating into unproductive arguments. On the other hand, weak or ineffective leadership will heighten tensions, stifle creativity, and lead to poor decision-making.

In Political Design, leadership is particularly important because the stakes are high. The decisions made by teams don't just affect the product, they have lasting consequences for society, consumers, and the economy. Strong leadership helps to navigate the complex political terrain of innovation, align the team's efforts with broader goals, and ensure diverse opinions are given the weight they deserve.

Chapter 4: Industry Culture

How Norms, Policies, and Trends Dictate Design
"The system shapes the solution before the designer picks up the pen."

In Political Design, the role of Industry Culture is essential to understand. While individual values and team dynamics influence how design takes shape within a specific team or organization, industry culture shapes the very environment in which those decisions are made. It's the backdrop, the set of unwritten rules, norms, policies, and trends that dictate how innovation happens or doesn't happen in any given industry.

Industry culture is like the invisible hand that guides the development of products, platforms, and technologies. Whether it's the established norms in the tech industry, the expectations of stakeholders in the financial sector, or the regulatory frameworks in healthcare, every industry has a culture that affects everything from what gets designed to how it's implemented and accepted by society.

The Power of Industry Norms and Policies

In any industry, the established standards of behavior and practice play a crucial role in shaping design. These norms are the **"rules of the game"** that designers and innovators follow, whether consciously or unconsciously. They often stem from both historical traditions and current best practices that have become widely accepted over time.

For example, in the tech industry, there are norms that prioritize speed over perfection, often leading to the rapid iteration of products and updates. These norms influence how teams approach the development of software, Apps, and platforms.

Similarly in the pharmaceuticals or financial industries, norms tend to emphasize compliance and risk management. This makes regulatory adherence a central part of the design process. These products must be designed with not only user experience in mind but also a strong focus on legal and policy standards.

Regulatory Frameworks: The Constraints of Innovation

The most explicit form of industry culture is the regulatory framework within which companies operate. Laws and regulations can have a profound impact on the way products are designed, implemented, and brought to market. These regulatory constraints can either encourage or stifle innovation.

For example, in the healthcare industry, regulations like the Health Insurance Portability and Accountability Act (HIPAA) in the U.S. dictate how health-related data can be collected, stored, and shared. Designers working on healthcare products must take these legal requirements into account when creating technologies that deal with patient data. Similarly, regulations around drug approvals dictate how pharmaceutical products are developed and tested, forcing design teams to navigate a complex web of compliance before they can bring a product to market.

Cultural Shifts and Their Impact on Design

Cultural shifts within society shaped by technological advancements, social movements, and economic changes have a profound impact on industry culture and Political Design. As society's values change, the expectations for design in every industry evolve accordingly.

This cultural shift in expectations has led companies to change the way they approach hiring, product design, and corporate policies. These shifts also impact the entire industry culture, pushing it toward more inclusive practices and equitable outcomes despite certain administration pushback.

Ethical design is also becoming an increasingly important topic. In the past, tech companies prioritized features and aesthetics without giving much thought to their social impact. Now, designers must account for issues like data ethics, digital addictions, and long-term consequences associated with their creations. Companies that fail to keep up with these cultural shifts risk falling out of touch with customers, investors, and valuable economic opportunities.

The Predefined Nature of Industry Culture

The culture of an industry may feel predefined, as if certain norms, policies, and practices are set in stone. However, it's important to recognize that industry culture constantly evolves. Technology, social movements, regulatory changes, and shifting economic landscapes all contribute to a continual redefinition of industry culture. What may be true today might not be true tomorrow. Old industries will adapt or die, new industries will emerge, and the landscape of Political Design will continue to evolve. Industry culture is predefined to be continually redefined.

Chapter 5: The Forces That Shape Political Design

Power. Influence. Access.
Breaking down the impact of power, influence, and access on decision-making and innovation.

In the world of Political Design, the processes that guide decision-making are rarely neutral. The forces of **power**, **influence**, and **access** constantly place pressure on what gets designed, how it gets designed, and who gets to be part of the process. These forces operate within the larger framework of individual values, team dynamics, and industry culture, but they remain crucial in shaping the very essence of what innovations come to life.

While these forces may seem abstract or intangible at first, they are important to understand if we are to grasp the intricacies of Political Design. Whether it's a tech giant rolling out a new product, a policy-maker shaping public health initiatives, or a team of designers considering their next big project, each of these forces plays a role in defining the possibilities and constraints that come with designing in an increasingly complex world.

Power: The Authority to Make Decisions

In the context of Political Design, **power** is the authority to create rules, impose regulations, and ultimately make decisions that guide what gets created and how it's implemented. This is the most direct and visible force at play in the design process. Power exists in various forms: corporate executives who steer product roadmaps, government regulators who enforce compliance, and international bodies that shape global standards.

The role of power is to set the boundaries within which design can take place. In some cases, it might even determine the outcomes by limiting options or enforcing specific conditions for design. Consider the role of regulatory bodies in industries like healthcare or finance. These authorities shape the design process not only by imposing legal constraints (such as data privacy laws) but also by determining the frameworks within which companies must operate.

Example:

In the tech industry, companies like Apple™ and Google™ often exercise their authority to create internal guidelines and design standards that not only influence the usability and appearance of products but also dictate user interactions. The Apple App Store™, for instance, enforces strict guidelines for App developers, dictating how Apps are designed, what data they can collect, and how they interact with the broader ecosystem.

Influence: The Subtle Force of Persuasion

While power enforces decisions, **influence** subtly shapes opinions and perspectives of design in ways that are often less visible. Influence can come from various sources: influential individuals, advocacy groups, public opinion, or media coverage. Unlike power, which is backed by authority, influence operates through dialogue, advocacy, and emotional appeal.

In the design process, influence works by persuading key decision-makers or the broader community to adopt a particular narrative or prioritize certain values. This could be in the form of advertising campaigns that sway consumer preferences, thought leadership that shapes industry trends, or even public interest groups advocating for more ethical design practices.

Example:

Consider the rise of privacy advocacy in the tech industry. In recent years, organizations like Electronic Frontier Foundation (EFF) and various privacy-conscious groups have persuaded tech giants like Facebook™ and Google™ to adopt stronger privacy features in their products. Through campaigns, legal actions, and public pressure, they've influenced the design of features like data encryption and consent-based data usage in social media platforms, which were previously neglected.

Access: The Force That Opens New Pathways

The third force, **access,** refers to the opportunity, resources, and avenues that allow innovation to take place. Without access, even the best ideas may never come to fruition. In Political Design, access is often shaped by factors such as proactiveness, market conditions, technological advancements, and global trends.

Access presents itself when external conditions align to create the potential for change. It could be a technological breakthrough that enables new kinds of products or services, or it might be an economic shift that opens up new markets for certain types of opportunities. In any case, access can be a powerful driver of design, enabling innovators to break away from traditional approaches and take risks that lead to groundbreaking solutions.

Example:

The increase of remote work technologies was another example of opportunity influencing design. As the COVID-19 pandemic accelerated the shift to working from home, technology companies rushed to adapt and leverage new communication and collaboration tools. The sudden need for virtual meeting platforms led to the rapid demand of tools like Zoom™ and Microsoft Teams™, which provide access for how people communicate and collaborate globally.

The Interplay of Power, Influence, and Access

Power, influence, and access do not work in isolation. However, they often work holistically and move each other to shape the outcomes of design.

For instance, power may create the rules and frameworks within which innovation happens, but influence can challenge those rules. Meanwhile, access conditions the opportunity for new solutions to emerge. These three solutions are often shaped by the ongoing conversations and persuasive efforts within individuals, teams, and industries.

The forces of power, influence, and access serve as crucial components that guide, restrict, or amplify innovation in Political Design. Understanding these components and how they interact is essential for innovators who aim to navigate the evolving landscape of technology, design, and humanity.

Chapter 6: Real-World Applications of Political Design

Case Studies of Political Design
"Understanding Political Design requires more than theory, it's about real experiences, decisions, and consequences."

Political Design shows itself differently across industries. The intersection of corporate culture, stakeholder influence, and project demands shape how design decisions are made and implemented. Below are explicit examples of industry types, real-life corporate scenarios, and lessons learned from navigating these challenging environments:

Marketing & Branding Agencies

- Fast-paced, reactive environments with aggressive deadlines.
- Overcommitment to client demands leads to long hours and burnout.
- Work-life balance is often sacrificed for rapid turnaround and client appeasement.

Tech Companies

- Agile methodology with iterative feedback loops.
- Feature deployment occurs over time, allowing for continuous improvement.
- Reflection periods help teams refine and optimize their solutions.

Financial Institutions

- Stringent regulatory policies and legal barriers impact design.
- Limited availability of key stakeholders slows progress.
- Compliance-driven decision-making restricts creative solutions.

Government Organizations

- Slower-paced, committee-driven decision-making.
- A consensus culture ensures selective inclusivity and may delay action.
- Fear of missing out on key recommendations leads to stakeholder over-involvement.

Non-Profit Organizations

- Purpose-driven with resource constraints.
- Long-term projects mixed with periodic urgent initiatives (e.g., fundraising).
- Requires balancing mission goals with operational feasibility.

Sports Industry

- Competitive, high-profile, and brand-driven.
- Frequent contract negotiations and endorsement management.
- Demands fast decision-making in a highly dynamic environment.

1 . Management Consulting

After completing a project in Seattle, I returned to the Chicago office only to be informed about a layoff during the 2009 recession. The consulting world operates on utilization rates and staying billable is key. One minute you're valued, the next, you're expendable.

Political Design:

Account Managers served as the main client contact due to limited stakeholder availability. Additionally, barriers between designers and primary clients slightly diluted the final experience.

Lesson Learned:

No matter how hard we work, there will always be more work to do. Prioritizing time wisely is crucial because time is a non-renewable resource.

2 . Retail Company

Despite thorough research and user testing, stakeholders consistently demanded last-minute changes. In one instance, our team was given an ultimatum: "Complete the solution before you leave today or you're fired."

Political Design:

To survive such high-stakes situations, we began creating multiple design options. A design that met our standards and another design that appeased leadership.

Lesson Learned:

Always have alternative design solutions prepared. Also establish career boundaries to determine which battles are worth fighting.

3 . News Agencies & Digital Media Company

A highly visible project was unexpectedly given a key stakeholder's name. The design team's expertise was questioned, and development constraints altered the final product design.

Political Design:

Not all projects are equal. Some projects carry more political weight and will influence decision-making more than others.

Lesson Learned:

Understand the power dynamics behind projects. What seems like a design challenge may actually be a political one.

4 . Open Source Tech Agency

Despite enjoying UX design, I transitioned into a Product Owner role. Over time, I realized my preference wasn't in managing product backlogs, but in designing user-centered experiences.

Political Design:

Feature prioritization is an ongoing negotiation between client demands, business requirements, developer constraints, and user experience recommendations.

Lesson Learned:

Health is wealth. No career role is worth compromising our well-being.

5 . Financial Institution

With multiple stakeholders from project sponsors to QA testers; obtaining design approvals can become a bureaucratic nightmare.

Political Design:

Filtering out unnecessary noise and prioritizing critical feedback is essential for progress.

Lesson Learned:

Confirm meeting attendees beforehand and tailor presentations accordingly to the audience attendance.

6 . Industrial Supply Company

Despite designing a point-of-sale system from ideation to implementation, key features were deprioritized into the "fast follow" backlog. This is a common corporate strategy to keep projects moving without committing to certain usability considerations.

Political Design:

When aggressive timelines are implemented, design trade-offs are inevitable.

Lesson Learned:

User research is invaluable. Engaging with end-users ensures design decisions align with customer needs, even if leadership prioritizes speed over quality.

7 . Marketing Agencies

Short-term design contracts were common. Clients often had their own research vendors, making in-house UX testing a conflict of interest.

Political Design:

In agency settings, marketing can often take precedence over UX. These design decisions are often driven by client preferences and not based on fundamental UX principles.

Lesson Learned:

Understanding industry priorities will sometimes trump business strategy and user experience best practices.

8 . Global E-Commerce Marketplace Company

Before a massive layoff, there was a 3 year term with the company; managing customer service, agent, and merchant experiences.

As a Product Design Manager, key project wins included improved post-purchase experiences, prompt agent response-time in chat sessions (e.g. CSAT), and merchant redemption transparency.

Political Design:

While developers focus on technical debt, UX teams must advocate for design debt. This ensures user experience isn't sacrificed for system functionality.

Lesson Learned:

When working on both customer-facing and internal systems, always balance business objectives with customer needs.

Key Takeaways:

Time is invaluable: prioritize it wisely.

Have alternative solutions: stakeholders may demand last-minute changes.

Recognize power structures: not all decisions are made rationally.

Protect your well-being: mental and physical health come first.

Understand industry priorities: sometimes, politics outweigh best practices.

By recognizing these patterns, we equip ourselves to navigate corporate realities without losing sight of the user experience and ethical integrity at the heart of design.

Chapter 7: Designing for Liberation

Navigate, challenge, and leverage Political Design for social impact.
"Power structures shape the world we navigate. The challenge is not in recognizing power structures, but in designing pathways toward altering them for positive change."

Start of the Ending: Rethinking the Foundations

Design is more than aesthetics and usability; it's a mechanism of influence. Every system, product, and experience is embedded with the biases, priorities, and power dynamics of the industries that create them. As we move forward, the question is not whether design will shape the future, but how intentionally we will shape design to be more revolutionary in nature.

This final chapter lays out 3 key principles and strategies for designing a future that prioritizes data, sustainability, and liberation:

Principle 1: Data is Relatively Subjective

Observation:

Data is relative and often treated as an objective truth, but it also reflects the biases of those who collect, interpret, and act on it. Questions to consider:

Whose data is prioritized? How does data influence or supplement decision-making? *Whose behavior patterns and experiences are observed?*

Application:

- Question how data is collected and who is represented.
- Demand transparency in algorithmic management and decision-making.
- Advocate for ethical AI principles and responsible data governance.

Example: Creating algorithms that filter candidates based on "cultural fit" may unintentionally reinforce racial, gender, and socioeconomic biases.

Principle 2: Global Sustainability is Non-Negotiable

Observation:

The future of design must consider not only people but the planet. The consequences of extractive business models disproportionately harm marginalized societies and accelerate global inequities.

Application:

- Prioritize environmentally sustainable design decisions.

- Advocate for ethical AI data center placement and regulations
- Support circular economic principles reducing waste and global warming factors.

Example: Tech companies promoting "innovation" while ignoring electronic waste (e-waste), toxic pollutants, and exploitation of non-renewable resources.

Principle 3: Design is a Tool for Liberation

Observation:

The most transformative design does not just create better products, it intentionally creates better systems and questions traditional methods.

Liberatory design boldly challenges power structures and imagines new possibilities for an evolving world. The sky is no longer the limit, but it's now limitless beyond the Karman line.

Application:

- Move from designing FOR communities to designing WITH communities.
- Use design as activism to amplify voices that are often silenced or ignored.
- Challenge exclusionary practices within the corporate environment.

Example: Leveraging liberatory design to create decentralized, transparent, and collaborative entities to impact social change.

Conclusion

Political Design is not an abstract concept; it's a real-world responsibility. Designers, strategists, leaders, and decision-makers hold the power to shape the future not just for profit, but for the planet, people, and future generations.

- Will we continue to reinforce existing power structures, or will we change them?
- Will we create for convenience, or will we design for disruption?
- Will we remain complicit in exclusion, or will we build systems that empower, collaborate, delegate, and liberate?

Design starts with curiosity, play, and passion. Over time, we're organically introduced to pain-points, policies, and politics, which intermingle with our innovative design experiences. This unique dynamic creates opportunities and space for problems to be solved. **No tension. No extension.**

Political Design has little to do with technical knowledge and implementation, but more about the soft skills of humanity: *communication, open-minded approaches to conflicts, likability, collaboration, presence, resilience, and empathy.*

Political Design

CONCLUSION

At its core, design is artistic problem solving;
with a jazz of interaction,
some blues of requirements,
with rock and roll politics;
seeking a sweet spot of harmony;
in a limited time frame;
before the product or service hits the door
for recipients to criticize, accept, or encore.

Appendix A

10 Steps for Starting a New Project from a Design Perspective

When initiating a new design project, it is essential to follow a structured approach that ensures clarity, efficiency, and alignment with project goals. The following ten steps provide a framework for navigating the design process effectively:

1 . What is the project about? (Executive Summary)

- Clearly define the project's purpose, scope, and intended outcomes.
- Summarize key objectives and goals to align all stakeholders.

2 . What stage is the project in? (Discovery, Design, Evaluation, Implementation, Testing)

- Determine the current phase of the project lifecycle to establish the appropriate design approach.
- Ensure the design team understands its role in the overall process.

3 . What are the specific expectations for the design team?

- Identify deliverables, responsibilities, and success metrics.
- Clarify whether the focus is on research, wireframing, prototyping,

usability testing, or full implementation.

4. What are the fundamental design challenges?

- Define the primary obstacles that the project aims to address.
- Consider technical limitations, user experience barriers, or business constraints.

5. Are there any existing features the project can leverage internally?
(e.g., non-customer-facing platforms)

- Evaluate whether internal tools, frameworks, or prior design assets can be repurposed.
- Avoid unnecessary duplication of effort by utilizing existing solutions.

6. Are there any similar features the project can leverage externally?
(e.g., competitive analysis)

- Conduct research on industry standards and competitor implementations.
- Identify best practices and potential areas for differentiation.

7. What constraints are present?

- Assess technological, budgetary, or time-based restrictions.
- Understand hardware/software limitations that may impact design feasibility.

8 . Who are the main decision-makers?

- Identify key stakeholders who influence project direction and approvals.
- Establish communication channels to ensure alignment.

9 . What are the main decision-makers' perspectives, opinions, or concerns?

- Gather insights from stakeholders regarding their expectations and potential reservations.
- Address conflicting viewpoints early to mitigate roadblocks.

10 . What's the expected deliverable date for design solutions?

- Establish a clear timeline for design milestones and reviews.
- Ensure deadlines align with broader project timelines and development cycles.

By following these ten steps, design teams can create a structured approach to new projects, ensuring efficiency, alignment, and successful execution.

Appendix B

5 Ways to Avoid Being Blindsided by a Company Layoff

Navigating job security in a constantly shifting work environment requires strategic foresight. The following five approaches help professionals remain prepared and resilient against unexpected layoffs:

1. Proactively Interview

- Maintain an ongoing dialogue with recruiters to stay informed about industry trends and salary expectations.
- Use informational interviews to refine communication skills and gain confidence in the hiring process.
- Treat early-stage interviews as practice rounds, preparing for high-stakes opportunities when they arise.
- Leverage your current employment as a bargaining tool when negotiating new offers.

"Well, my mama always said there was nothin' wrong with talkin'." – Ray (2004)

In an "at-will" employment culture, staying open to new opportunities ensures professional stability.

2. **Stash Funding**
- Secure financial independence by maintaining an emergency fund.
- Avoid making rushed career decisions out of financial desperation.
- Allocate resources strategically to provide time for evaluating career options.
- Recognize that while companies provide compensation, financial security is a personal responsibility.

"It's nothing personal... It's just business."

Building financial resilience ensures flexibility in navigating career transitions.

3. **Maintain Relationships**

- Keep professional connections active, as the industry network remains interconnected.
- Engage former colleagues who can provide referrals and insights into new opportunities.
- Leverage digital platforms to sustain relationships and industry visibility.
- Understand that professional networks often accelerate the hiring process.

"Relationships provide access, save time, and create opportunities."

Cultivating authentic relationships strengthens career mobility.

4. **Export Assets**

- Regularly document and back up key projects and work samples (within legal and contractual boundaries).

- Anticipate sudden loss of access to company-issued devices and platforms.
- Create independent records of contributions and accomplishments.
- Recognize that corporate equipment is loaned property, subject to immediate revocation.

"One of the challenges of being blindsided by a layoff is the restricted access that may occur after the initial announcement."

Proactive asset management ensures continued professional leverage.

5. **Eeny, Meeny, Miny, Moe – Have Options**

- Diversify income streams beyond a single employer.
- Explore investments, freelance work, or passive income sources to reduce financial dependency.
- Recognize that employment benefits can be altered or removed at any time.
- Align current earnings with long-term wealth-building strategies.

"Eeny, Meeny, Miny, Moe! Having options is the way to go!!"

A well-balanced professional and financial portfolio safeguards against unexpected setbacks.

Appendix C

Tell-Tale Signs It's Time to Leave a Company

Recognizing when to move on from a company can be difficult, but certain patterns indicate when a workplace is no longer serving your professional growth, mental well-being, or financial interests. The following signs suggest it may be time to consider new opportunities:

1. **Your Role No Longer Aligns with Your Career Goals**

   - The responsibilities of your job have significantly shifted from what you were originally hired to do.
   - Your skills and ambitions are no longer being developed in a meaningful way.
   - You find yourself on a career path that does not align with your long-term goals.

2. **Your Ideas Are Initially Rejected, Then Later Adopted Without Credit**

   - A proposed solution is dismissed, only to be reintroduced later by those who originally rejected it.
   - Your contributions are not acknowledged, despite proving to be valuable.
   - You feel undervalued as leadership fails to recognize your strategic insights.

3. **You Are Laid Off with a Severance Package**

- A sudden restructuring results in your position being eliminated.
- You are given a severance package and an exit discussion—confirming the company's direction no longer includes you.
- The decision is final, signaling that it's time to transition.

4. **The Thought of Going to Work Makes You Physically Sick**
- You experience anxiety, stress, or dread at the thought of starting your workday.
- The workplace has become a source of mental and emotional strain.
- Your job is negatively impacting your overall well-being.

5. **You Are Forced Into Survival Mode Instead of Being Supported**

- You find yourself constantly creating workarounds instead of implementing optimal solutions.
- Resources and project support are lacking, forcing you to cut corners.
- The company prioritizes short-term fixes over long-term success.

6. **You Resent New Team Members for Their Ideas**
- You feel frustration when new hires introduce innovative solutions.
- Instead of welcoming fresh perspectives, you see them as unnecessary work that will not be implemented.
- Your passion for problem-solving has diminished due to systemic project failures.

7. **Your Intellectual Property Is No Longer Your Own**

- Your creative ideas and designs are absorbed into the company without proper compensation.
- You receive no equity or long-term stake in the success of your work.
- Innovation is no longer personally rewarding.

8. **There Is No Longer Representation or Advocacy at the Executive Level**

- Leadership no longer includes individuals who understand or support your skill set.
- Your department or team lacks internal advocates for necessary resources and growth opportunities.
- The company's priorities have shifted away from your area of expertise.

9. **You Are Ready to Be an Entrepreneur**

- You no longer want to work under someone else's vision.
- You are prepared to own your ideas, take financial risks, and build something independently.
- You seek to create a legacy that benefits your family and future generations.

Recognizing these tell-tale signs allows professionals to make informed decisions about their careers. Whether it's moving to a new company or pursuing entrepreneurship, understanding when to leave ensures that career transitions happen on your terms.

Appendix D

Is DEI in the DNA?

Is Diversity, Equity, and Inclusion (DEI) truly embedded in a company's DNA, or does it disappear when financial pressures arise?

During times of downsizing, reorganization, hiring freezes, and staff reductions, DEI initiatives are often among the first to be deprioritized or eliminated. This raises an important question: Was DEI ever truly integrated into the company's core values, or was it merely a temporary initiative subject to financial convenience?

For DEI to be effective and enduring, it must go beyond being an initiative housed under Human Resources (HR) or Marketing. Instead, it should be elevated to the C-Suite level, ensuring that diversity, equity, and inclusion are fundamental to corporate decision-making and long-term strategy.

When DEI becomes a foundational pillar of an organization, it fosters an inclusive environment that thrives even in times of economic uncertainty. This strengthens company culture, drives innovation, and enhances operational resilience.

So, how can organizations embed DEI into their core structure and prevent it from fading away?

10 Ways to Prevent DEI from Disappearing in the Workplace

1 . Request Demographic Information from HR (If Permitted by Company Policy)

- Obtain a clear picture of the company's workforce composition.
- Use this data to identify representation gaps and establish appropriate Employee Resource Groups (ERGs).

2 . Identify DEI Leadership and Team Membership

- Determine who is responsible for DEI efforts and establish accountability at different levels of the organization.

3 . Host Introduction Sessions for DEI Groups

- Set clear meeting frequencies, objectives, and agendas to ensure structured engagement.

4 . Gather Employee Feedback Using Multiple Methods

- Use surveys, focus groups, breakout sessions, and one-on-one interviews to assess employee experiences and perspectives.

5 . Prioritize and Categorize DEI Findings

- Organize feedback into actionable themes, highlighting areas needing immediate attention.

6 . Share Executive Summaries of DEI Feedback

- Provide transparent updates to ERGs, leadership, and employees to build trust and accountability.

7 . Locate Internal Allies and Champions

- Secure sponsorship from C-Suite executives and senior leaders to advocate for DEI initiatives.

8 . Develop and Uphold DEI Bylaws

- Establish formal governance structures that define DEI policies, responsibilities, and long-term objectives.

9 . Encourage Volunteering, Networking, and Mentorship

- Foster community engagement and career development opportunities for employees from underrepresented groups.

10 . Align HR and ERG Leaders for Recruitment Strategies

- Strengthen DEI-driven hiring pipelines by collaborating with educational institutions, local communities, and faith-based organizations.

By implementing these strategies, organizations can ensure that DEI is not just a temporary initiative but an integral part of the company's DNA. This commitment to DEI fosters a culture where innovation, collaboration, and various perspectives drive long-term success.

About the Author

Lloyd Hervey II brings 15+ years of UX strategy, agile leadership, and industry experience to deliver impactful, user-centered solutions. With an MS in Human-Computer Interaction (HCI) from DePaul University and a Certified Scrum Product Owner (CSPO), Lloyd consults for clients and drives measurable results through creative insight, collaboration, and social impact.

You can connect with me on:

🔗 https://www.linkedin.com/in/lloydherveyii

www.ingramcontent.com/pod-product-compliance
Lightning Source LLC
Chambersburg PA
CBHW040931210326
41597CB00030B/5259